Elena Bukvareva

The Summary of the Principle of Optimal Diversity of Biosystems

Elena Bukvareva

The Summary of the Principle of Optimal Diversity of Biosystems

LAP LAMBERT Academic Publishing

Impressum / Imprint

Bibliografische Information der Deutschen Nationalbibliothek: Die Deutsche Nationalbibliothek verzeichnet diese Publikation in der Deutschen Nationalbibliografie; detaillierte bibliografische Daten sind im Internet über http://dnb.d-nb.de abrufbar.

Alle in diesem Buch genannten Marken und Produktnamen unterliegen warenzeichen-, marken- oder patentrechtlichem Schutz bzw. sind Warenzeichen oder eingetragene Warenzeichen der jeweiligen Inhaber. Die Wiedergabe von Marken, Produktnamen, Gebrauchsnamen, Handelsnamen, Warenbezeichnungen u.s.w. in diesem Werk berechtigt auch ohne besondere Kennzeichnung nicht zu der Annahme, dass solche Namen im Sinne der Warenzeichen- und Markenschutzgesetzgebung als frei zu betrachten wären und daher von jedermann benutzt werden dürften.

Bibliographic information published by the Deutsche Nationalbibliothek: The Deutsche Nationalbibliothek lists this publication in the Deutsche Nationalbibliografie; detailed bibliographic data are available in the Internet at http://dnb.d-nb.de.

Any brand names and product names mentioned in this book are subject to trademark, brand or patent protection and are trademarks or registered trademarks of their respective holders. The use of brand names, product names, common names, trade names, product descriptions etc. even without a particular marking in this works is in no way to be construed to mean that such names may be regarded as unrestricted in respect of trademark and brand protection legislation and could thus be used by anyone.

Coverbild / Cover image: www.ingimage.com

Verlag / Publisher:
LAP LAMBERT Academic Publishing
ist ein Imprint der / is a trademark of
OmniScriptum GmbH & Co. KG
Heinrich-Böcking-Str. 6-8, 66121 Saarbrücken, Deutschland / Germany
Email: info@lap-publishing.com

Herstellung: siehe letzte Seite /
Printed at: see last page
ISBN: 978-3-659-59891-3

Copyright © 2014 OmniScriptum GmbH & Co. KG
Alle Rechte vorbehalten. / All rights reserved. Saarbrücken 2014

This paper is dedicated to Gleb Aleshchenko, my long-term friend and collaborator. Without him, this work would not be possible

The paper summarizes the principle of optimum diversity of biosystems, which suggests that biodiversity is a parameter to be optimized. In fact, diversity is considered as a major adaptation of biosystems to environmental conditions. Biosystems with the optimal diversity have maximum efficiency and probability of survival. Paper discusses the diversity of two hierarchical levels - population (phenotypic diversity within a population) and coenotic (number of species in community). It is shown that the optimal values of diversity are determined by the amount of a resource in the environment, the degree of environmental stability and by the evolutionary level of organisms. The adaptation of biosystems to environmental conditions occurs through the optimization of diversity at the population and community levels during their interaction. Optimal values of species diversity increase in more stable and "rich" environments, while optimal values of intrapopulation diversity decrease in more stable environments and is independent of the intensity of resource flow. These opposite reactions allow us to make an assumption of the different roles of intrapopulation diversity and species diversity in a fluctuating environment: intrapopulation diversity is the basis of adapta-tion to environmental instability, while species diversity enables a community to use resources to the maximum and effectively.

The predictions of the principle of optimal diversity does not contradict the basic array of empirical data, and in some cases they are confirmed. This allowes us to accept the principle of optimal diversity as a working hypothesis and put forward on this basis specific hypotheses about mechanisms of diversity optimisation in the ecological, microevolutionary and evolutionary processes.

The final section of the paper discusses the findings from the principle of optimal biodiversity for strategies for sustainable environmental management and biodiversity conservation.

The author expresses his deep gratitude to Gleb Aleshchenko for long-term high-quality work on the creation of mathematical models of optimal biodiversity and their software implementation, as well as the Alexey Severtsov for their valuable comments and additions to the preparation of the dissertation.

This work was supported by the program of the Presidium of RAS "Wildlife: current status and problems of development."

Content

Introduction

Modern large-scale shrinking of natural ecosystems and destruction of biodiversity lead to weakening of ecosystem functions and mechanisms of natural environment regulation. Damage from that becomes an important factor in the economy and safety. Concern about the possible consequences of diversity loss has led to a rapid growth in the last 20 years the number of studies of the relationship between biodiversity and ecosystem functioning. The obtained results proved that biodiversity is an important factor which determines stability and functioning of ecosystems.

Optimization principles may give a considerable benefits in the investigations of interconnections between ecosystem properties and diversity. Such principles have got wide distribution in physiology, biochemistry, embryology, evolution theory, population dynamics, and ecology. However, in the field of biodiversity research, the capabilities of this method have not been used in full measure. Some works consider the maximum diversity, but not optimal. For example the "entropy extreme principle" for communities implies the maximization of community complexity at fixed volumes of resource consumption by different species (Levich, Alekseyev, 1997). The other example is the "principle of maximum diversity of biomass distribution" in the population (Lurie et al., 1983; Wagensberg, Valls, 1987).

This paper presents the summary of the optimization approach, suggesting that diversity of elements of a biosystem is the adjustable variable, which is optimized and allows biosystems to maximize their effectiveness and vitality (Букварева, Алещенко, 2005; more detailed description of the principle is in the book: Букварева, Алещенко, 2013). Y. P. Altukhov previously had proposed the idea of existence of the optimal gene diversity within populations (Алтухов Ю.П., 2003). The presented in this paper hypothesis combines for the first time optimization of diversity of two hierarchical levels - populations and ecological communities. It can be extended to other hierarchical levels of biosystems (Букварева, Алещенко, 1997, 2013), but in this paper we will consider only levels of populations and communities, which further for brevity called "biosystems".

1. The general formulation of the optimal biodiversity principle

The optimal biodiversity principle is based on the suggestion that the diversity of elements of a biosystem is related to its viability (survival probability). The viability is maximal at the optimal level of diversity (Fig. 1). It is also possible existence of critical levels of diversity, in which the biosystem becomes unviable. The biosystem is trying to reach a state with maximum viability and optimal diversity (V*, D*), but it is quite difficult to achieve it. The diversity of undisturbed natural populations and communities may be the closest to the optimal values. An artificial decrease or increase of inner biosystem diversity leads to a decrease of its viability.

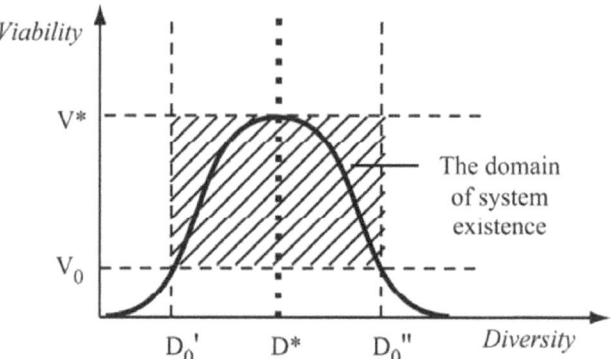

Figure 1. *Optimal value of diversity (D*) corresponds to maximum biosystem viability (V*). V_0, a critical value of viability; D_0, a critical value of diversity; the shaded area is a domain of system existence.*

2. Models of the optimal biodiversity

2.1. The model of phenotypic diversity of the population

The model presents an isolated population which exists in some environment (Алещенко, Букварева, 1991). The formal descriptions of the models are presented in the Supplement. Environment is characterized by the degree of intensity and stability of available resource flow. There is some environmental parameter that can be interpreted as a characteristic of resource (e.g. light wave length, size of prey, etc.) or as an environmental factor that allows resource consumption (e.g. temperature, humidity, etc.). At each moment of time, some value of this parameter is realized. The dispersion of the distribution of its values σ^R defines the degree of environmental instability.

Population dynamics is described by simple equations: the death rate is set by exponential dependence with a constant mortality; reproduction is modeled by a logistic function.

The population consists of various phenotypes. Phenotype characteristic is the ability of individuals to reproduce in a given environmental conditions (Fig. 2). At each moment of time, the realized value of environmental factor f* corresponds to a certain phenotype, for which the realized conditions are the most favorable. At this moment, a group of phenotypes breeds around it. The value of dispersion of distribution of breeding phenotypes σ^A (black bars in Fig. 2) can be interpreted as the width of the zone of individual tolerance. The value of dispersion of distribution of their offspring σ^B (shaded bars) serves as an index of diversity reproduced by the population at each step of its development. These two parameters - diversity of breeding phenotypes σ^A and diversity of offspring phenotypes σ^B - formed in the course of the experiment the total phenotypic diversity of the population σ^X (white bars).

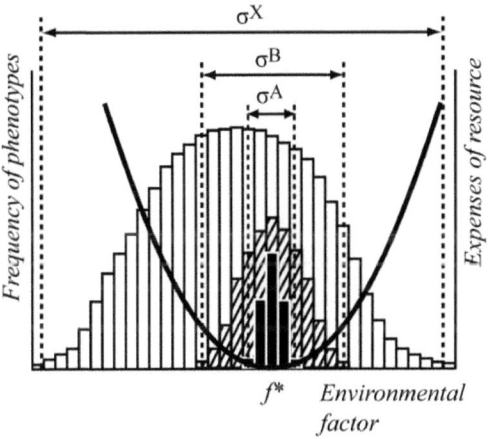

Figure 2. *Phenotypic diversity in population and resource spending by phenotypes. f*, the value of environmental parameter realized at a given moment of time; black bars, phenotypes breeding at a given moment; shaded bars, offspring of the breeding phenotypes; white bars, existing phenotypes; black curve, resource spending by phenotype f*.*

Individuals spend some resource for existence and reproduction. The farther the realized environmental parameter is from the optimal value for a given phenotype, the greater the resource spending by this phenotype (Fig. 2).

During computer experiments, populations die out or reach some stationary quantity with definite phenotype diversity and with some level of resource consumption. Dispersion of population size, i.e. the measure of its fluctuations σ^N for long time intervals was used as index for stability of the population.

The width of the stationary distribution of phenotypes σ^X (white bars in Fig. 2) is an index of intrapopulation phenotypic diversity which can be compared with the width of the ecological niche of the population. Diversity of offspring can be compared with between-individual component of ecological niche, and diversity of breeding phenotypes (individual tolerance zone width) - with within-individual component. Links between genetic and phenotypic diversity are not considered in the model (they are present in a hidden form in the parameters σ^A, σ^B, r_{max}, d).

The optimality criterion is the maximum size (biomass) of population at a fixed volume of available resource. This task is equivalent to the minimization of resource spending per individual at a fixed population size (biomass). Phenotypic diversity, in which the maximum population size is reached, is the optimal diversity.

2.2. Two-level hierarchical model "population - community" without divergence of ecological niches

The community consists of populations which share the available resources, thus it is a community of one trophic level (Aleshchenko, Bukvareva, 2010). Thus, the lower hierarchical level is represented populations, the top level - the community. On the population level is used the model described earlier. The number of populations in the community is considered as the number of species. All populations are identical in their parameters, phenomenon of dominance is not considered. Populations are identified not on the basis of ecological niches (which are the same for all populations), but on the basis of unity of the process of reproduction (all phenotypes reproduce only inside their populations).

The optimization criterion for the community is the maximum of total quantity of individuals of all populations (or total biomass) at a fixed volume of available resource. This task is equivalent to the minimization of resource spending under the condition of full consumption of the available resource.

Optimal values of diversity are adjusted during iterative interaction of the two hierarchical levels by the following steps:

- each population consumes all the resources allocated to it and tries to reach the maximum size (biomass) by setting its inner diversity at the optimal level;

- the values of population size chosen at the lower level are transferred upward to a level of community;

- the upper level in view of these values defines the number of populations (number of species) at which the total quantity of individuals (total biomass) is maximum;

- available resource is divided into parts according to the selected number of populations, and each population gets the corresponding proportion of the total resource;

- recurrence of the first step: populations solve their optimization problem consuming the resource allocated to them, etc.

As a result of multiple iterations, the final values of optimal diversity are established in populations and community.

2.3. The model of community with the possibility of divergence of ecological niches

Populations have the ability to disperse on the axis of environmental parameter that is to separate niches (Bukvareva, Aleshchenko, 2013). For simplicity we consider as an environmental parameter some characteristic of the resource. The axis of resource parameter consists of a number of cells and forms a ring to avoid boundary effects. At every moment of time resource flows in some random cells (Fig. 3).

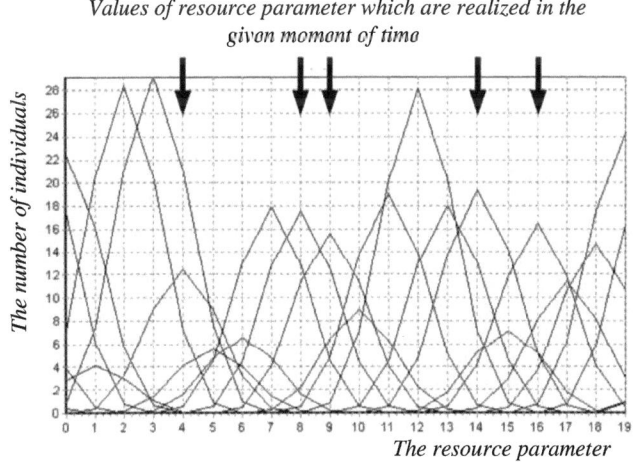

Figure 3. Distribution of species on the axis of the resource parameter at some moment of time. Each "bell" represents a population (option model with 20 cells).

7

The degree of environmental stability is determined by the number of cells with resource at each step – the more stable the environment has resource in the more number of cells. A number of populations (species) live in this environment. Each population consists of different phenotypes. As in the first model, the phenotypic feature - the ability to reproduce under certain environmental conditions, i.e. in a specific cell. Mechanisms of reproduction and distribution of phenotypes are similar to those in the first model. Initially, species are placed in each cell of the resource axis. After several generations the community passes to the steady state or all species die. The width of species phenotypic distribution in the steady state is interpreted as the width of its ecological niche. Niches of species may overlap.

3. The results of modeling

3.2. Optimization criteria used in the models correspond to the maximum efficiency of biosystems

The same optimization criteria are used at the lower (population) and the upper (community) levels of the models. In fact, these criteria may be reduced to one - the minimum resource cost of an individual or of a unit of biomass. That is, populations and communities with optimum values of diversity are the maximum effective. Such optimization criterion is quite plausible, since it is directly related to the viability of biosystems. Decrease in resource costs to maintain the number (or biomass) of a population or a community will increase the likelihood of their survival (viability).

Thus, the optimal values of diversity correspond to maximum efficiency and viability of populations and communities.

3.3. There are ranges of diversity, in which populations are stable

There is a range of values of offspring diversity σ^B and general phenotypic diversity σ^X at which the population is stable in the given environment. When the population leaves this range for a decrease or increase, it becomes unstable. Fig. 4 shows the dependence of the dispersion of population size σ^N on the offspring diversity σ^B (black dots) in the steady state. There is a distinct area of small values of σ^N, which corresponds to the zone of population stability.

Since a diversity of offspring phenotypes σ^B is linearly related to total phenotypic diversity σ^X, all findings can be generalized to the total phenotypic diversity of the population (an example is shown in Fig. 6).

The causes of loss of population stability at extremely low phenotypic diversity are obvious - the probability of convenient environmental conditions decreases. The loss of stability at high diversity is less obvious: it occurs because each phenotype class has too few individuals and so the probability of population extinction increases. In less stable environments, the stability range is reduced owing to the populations with low indexes of birth rate, high mortality and low phenotypic diversity.

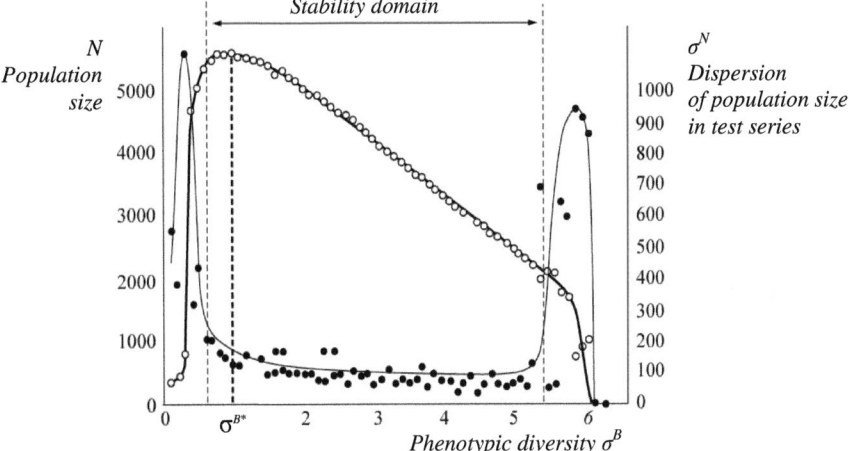

Figure 4. *Dependence of population size N (white circles) and its dispersion σ^N (black circles) on offspring phenotypic diversity σ^B. σ^{B*}, optimal phenotypic diversity.*

3.1. The optimum values exist both for intrapopulation and species diversity

The model experiments reveal the existence of optimal values of phenotype diversity which correspond to the maximum population size or biomass (or minimum resource costs). The example of emergence of optimal phenotypic diversity using the criterion of maximum population size is shown in Fig. 4. The example using the criterion of minimum resource costs is shown in Fig. 6. That is, any deviation from the optimal phenotypic diversity leads to decrease in the average population size or to an increase in resource costs.

It is interesting to note that the optimal values of diversity are close to the lower border of population stability. If we suppose that natural populations have phenotypic diversity close to optimal values, this result emphasizes the danger of reduction of intrapopulation diversity.

The optimal value of the number of species in a community (i.e. number of populations) also exist and correspond to the maximum total quantity or biomass of all populations (or minimum resource costs depending on the chosen optimization criterion).

Optimal values of intrapopulation and species diversity arise in the interaction of population and community levels under specific environmental conditions.

3.4. Optimum diversity values depend on the environmental parameters – the amount of resource and stability of environment. Populations and communities respond to changes in the stability in the opposite way.

The optimal values of the phenotypic diversity of offspring (σ^{B*}) and total phenotypic diversity (σ^{X*}) increase in more unstable environments. In other words, the population in a less stable environment need higher diversity to reach maximum size. At the same time in unstable environment the maximum population size is less than in stable conditions (Fig. 5).

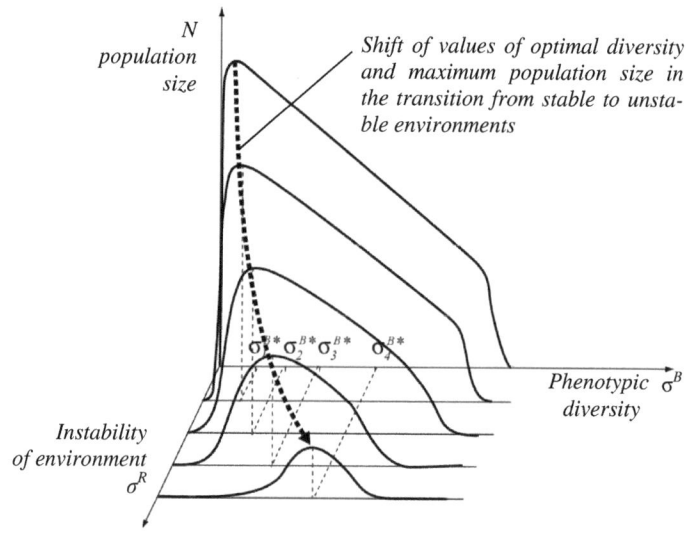

Figure 5. *Optimal values of phenotypic diversity (σ^{B*}) and population size in environments with different degree of instability (σ^{R}).*

Using the criterion of minimum resource costs optimal values of phenotypic diversity change similarly, with costs rising at less stable environments (Fig. 6).

The optimal number of species (populations) in the community changes in the opposite way – in less stable conditions the optimal number of species decreases. However, the total community biomass decreases as well as in populations.

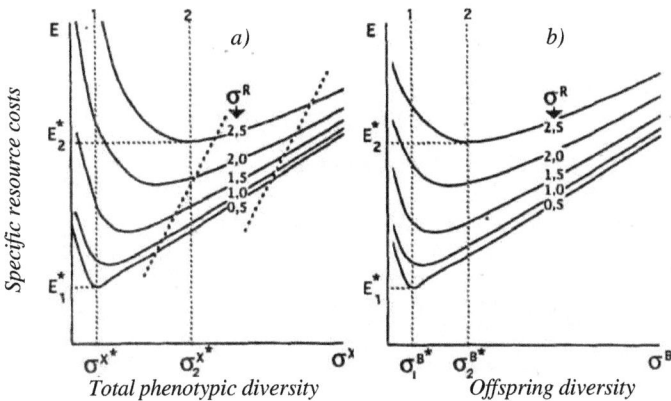

Figure 6. Specific expenses of resource in dependence on the total phenotypic diversity (a) and offspring diversity (b). σ^R shows the degree of environmental instability. The population "1" is adapted to more stable environment ($\sigma^R = 0,5$) and has optimal diversity values σ_1^{X*} and σ_1^{B*}. The population "2" is adapted to less stable environment ($\sigma^R = 2,5$) and has optimal diversity values σ_2^{X*} and σ_2^{B*}. E_1* and E_2* - minimum expenses of resource by populations "1" and "2", respectively.

In the model with the divergence of ecological niches the optimal values of diversity depend on the environmental stability in the same manner: in less stable environments optimal value of intrapopulation diversity (width of niches) increases, and the optimal number of species decreases (Fig. 7).

Thus, in less stable environments the optimal diversity in populations increases, optimal diversity in communities decreases, the effectiveness of populations and communities decreases (using the same flow of resources they can maintain a lower biomass). In more stable environments the opposite is true: the optimal intrapopulation diversity decreases, the optimal number of species increases, the effectiveness of populations and communities increases. The opposite reaction of population and coenotic levels on changes in the degree of environmental stability leads to the assumption of the different role of diversity at these two levels: intrapopulation diversity is the basis for the adaptation of populations and communities to instability of environment and species diversity allows the community to use resources more efficiently by separation of niches.

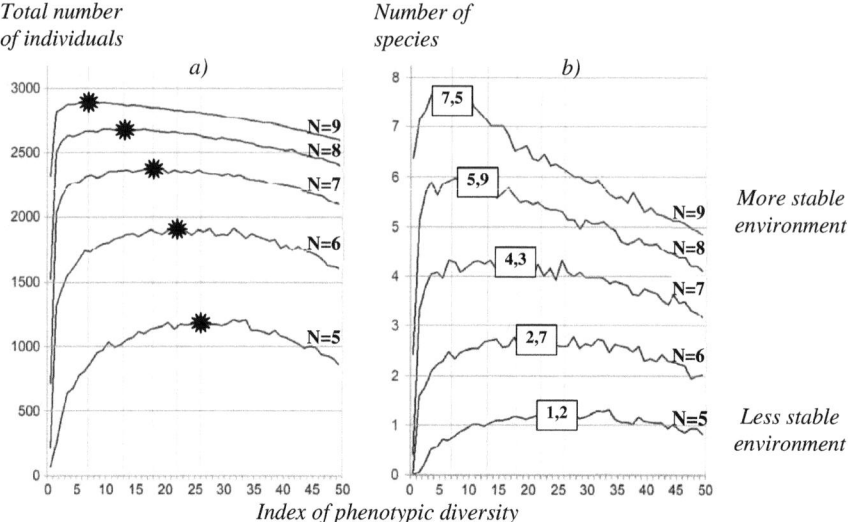

Figure 7. Dependence of the optimal diversity values on environmental instability in the model with divergence of ecological niches. Increase the optimal phenotypic diversity (asterisks in Fig. a) and decrease the optimal number of species (numbers in frames in Fig. b) in less stable environments. N is number of cells with the resource influx which corresponds with stability of environment.

The amount of resource does not affect the optimal values of intrapopulation diversity and increases the optimal number of species in the community.

3.5. Optimum values of diversity depend on the parameters of model populations

Optimum values of diversity depend on the following population parameters (offspring diversity was considered as a degree of freedom in the computational experiment):

- diversity of phenotypes which breed at each moment, which can be considered as the width of the zone of ecological tolerance of individuals;

- the maximum population growth rate and mortality rate;

- the resource costs of phenotypes and the function of its growth depending on how far the realized value of environmental factor is from the optimal value for current phenotype.

Progressive changes in any of these parameters (widening of the individual tolerance zone, increase in the maximum rate of population growth, decrease in mortality or resource costs) lead to qualitatively identical result – decrease in the optimal values of intrapopulation diversity, increase in the effectiveness of populations (Fig. 8), and therefore, increase in the optimal values of species diversity and raise of the effectiveness of communities.

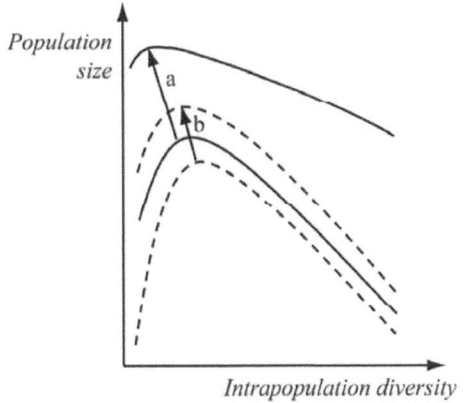

Figure 8. Changes in optimal values of intrapopulation diversity and effectiveness of populations: a) at increase in the maximum population growth rate; b) at increase in individual's ecological tolerance.

All the above changes produce the same effect on position of optimal diversity values as stabilization of environment. Thus, except of adjustment of the niche width, populations have different ways to compensate environmental instability: to increase population growth rate, to decrease mortality or to broaden the zone of individual tolerance. This mechanism can work on the level of one population within its adaptive capabilities, and on the level of community due to the change in species composition, for example, shifting between K- and r-strategists or between specialists and generalists.

Progressive changes of any of the above population parameters while maintaining constant the other characteristics can be interpreted as an increase in evolutionary level of organisms (e.g. decrease in mortality or broadening of individual tolerance). Thus, it can be assumed that raising the evolutionary level organisms leads to higher optimal values of species diversity and lower optimal intrapopulation diversity.

3.6. Expected values of species and intrapopulation diversity in undisturbed communities

Thus, the general conclusions about changes of parameters of model populations and communities in relation to environmental conditions are as follows:

- optimal values of intrapopulation diversity decrease in more stable environments and do not depend on the amount of resource;

- maximum sizes of populations are bigger in more stable environments and in more "rich" environments (with more resources);

- optimal numbers of species and the corresponding values of the total community biomass are higher in more stable environments or in more "rich" environments.

These results suggest that natural communities which are adapted to "rich" and stable environments consist of a large number of species with low intrapopulation diversity (specialists), while communities which are adapted to "poor" unstable environments consist of a small number of species with high intrapopulation diversity (generalists). In "rich" unstable and "poor" stable environments, we may expect some intermediate values of diversity (Fig. 9).

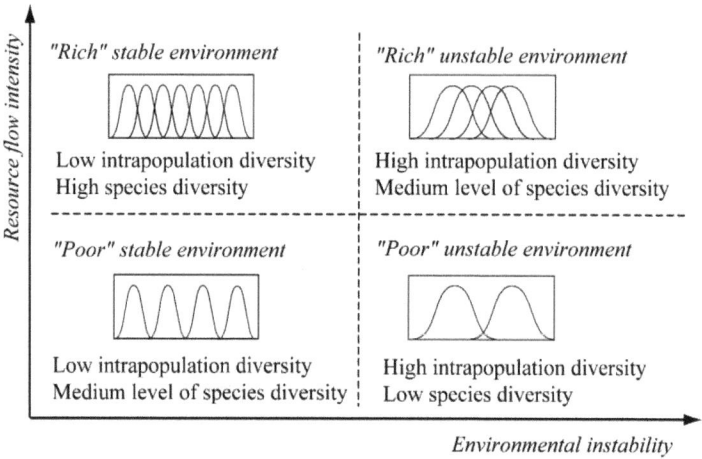

Figure 9. *Expected values of species and intrapopulation diversity in communities which are adapted to different environments.*

These predictions are made for undisturbed natural systems which exist in ecological equilibrium in a historically typical environment, that corresponds to a climax communities. Populations and communities, disturbed by people (exploitation, pollution, etc.), or existing under frequent natural disturbances (e.g., in areas of unstable river channels, landslides, storm damages, etc.), are far from optimal diversity values. Isolated communities with deficient species richness can also be attributed to suboptimal systems.

4. The qualitative verification of the optimal biodiversity principle

Verification of the models of optimal diversity biosystems at this stage of research can be carried out only qualitatively. Quantitative verification task was not intended.

319 individual studies and reviews and 15 meta-analyses, including data over 1000 experiments and observations were analyzed (Букварева, Алещенко, 2013). The spectrum of considered biosystems include populations (microorganisms, plants, invertebrates and vertebrates), marine, freshwater and terrestrial communities (e.g. soil communities, communities of marine and freshwater fish, invertebrates, algae, communities of herbaceous plants, forest, tundra and desert communities, etc.). These examples show that the results of experiments and surveys of natural communities don't contradict the following provisions of the optimal biodiversity principle:

- increase in the amount of resource leads to an increase in the number of species;

- increase in the degree of environmental instability leads to an increase in intrapopulation diversity (ecological niche width) and reduction of the number of species;

- reduction of the number of species may be accompanied by an increase in intrapopulation diversity (ecological niche width);

- increase the evolutionary level of organisms may lead to an increase in the optimal number of species and reduction of between-individual component of ecological niches, and vice versa;

- the maximal indexes of functioning are observed at the optimal intrapopulation and species diversity (not minimal or maximal);

- populations are most stable at the optimal intrapopulation diversity, with its reduction or increase, they lose stability.

Available in the world literature empirical data on biodiversity depending on environmental factors and biosystem functioning depending on biodiversity, do not contradict the provisions of the optimum biodiversity principle, and in some cases confirm them. This allows us to consider the principle as a general working hypothesis and on this basis to formulate specific hypotheses.

5. Hypothesis about processes and mechanisms of biodiversity optimization

It is convenient to consider processes of biodiversity optimization in terms of "license-niche concept" of Starobogatov and Levchenko (Левченко, Старобогатов,

1990; Старобогатов, Левченко, 1993). Licenses are regarded as sets of conditions, which ecosystem provides for populations (instead of logically contradictory notion of "empty niches"). Based on the optimal biodiversity principle we can assume that the "geometry" of space of licenses is determined by the optimal values of intrapopulation diversity (license width) and optimal number of species (number of licenses). In this discussion we examine only width and the total number of licenses but not functional performance and position of licenses in ecological space. Parameters of space of licenses are determined by the environment (the amount of resource and the degree of stability) and by the characteristics of species.

5.1.   The cause of optimization processes is mismatch between the existing community structure and optimal diversity parameters

Optimization of diversity occurs in cases when the existing number of species and width of ecological niches do not correspond to the optimal diversity parameters (optimal geometry of license space). The main groups of such causes are as follows (some combinations of them are most likely).

1 – Changes of abiotic conditions (climate, geological processes, anthropogenic impacts, including pollution and eutrophication, etc.) or biotic environment (introduction of alien species, anthropogenic extinctions, etc.) that leads to a shift previously achieved balance between the realized and optimal diversity.

2 - Changes in the environment in the course of community succession, which cause changes optimal parameters of licensing and niche space (see below).

3 - The discrepancy between the characteristics of the regional species pool and optimal parameters of license space, including cases of lack of species in the isolated habitats (islands, lakes, etc.).

4 - Change the characteristics of species in the course of their evolution, which also alters the optimal values of intrapopulation and species diversity.

In the first two cases, the optimal values can be achieved in the ecological processes, in the third case optimization can go through microevolutionary processes, the fourth group of causes relates to the optimization of biodiversity in the course of evolutionary processes (see below).

On the base of the results of modeling we can identify the main directions of diversity optimization when changes in the environment (Table 1).

Table 1. *Discrepancy between the realized and the optimum diversity values as a result of changes in the environment and the directions of optimization of diversity.*

Changes in the environment		Discrepancy between the realized and the optimum diversity values	Directions of diversity optimization	
Stability	Amount of resource		Intrapopulation diversity	Number of species
More stable	No change or increase	Niches are wider than optimal The number of species is less than optimal	Decrease (narrowing niches, specialization)	Increase (immigration, speciation)
	Decrease	Niches are wider than optimal The number of species is optimal	- \\ -	Unchanged
	Strong decrease	Niches are wider than optimal The number of species is more than optimal	- \\ -	Decrease (local extinction)
No change	Increase	Unchanged	Unchanged	Increase (immigration, speciation)
	Decrease	Niches have optimum width The number of species is more than optimal	Unchanged	Decrease (local extinction)
Less stable	No change or decrease	Niches are narrower than optimal The number of species is more than optimal	Increase (expanding niches, despecialization)	Decrease (local extinction)
	Increase	Niches are narrower than optimal The number of species is optimal	- \\ -	Unchanged
	Strong increase	Niches are narrower than optimal The number of species is less than optimal	- \\ -	Increase (immigration species, speciation)

5.2. Ecological, microevolutionary and evolutionary aspects of optimization

As noted above, the optimal values of diversity can be considered as the optimal number and width of licenses. But this is only the basis for the formation of diversity in real life. Biosystems can achieve or not optimal diversity during ecological, microevolutionary and evolutionary processes (Букварева, Алещенко, 2010). Duration of these processes is conditional and they may occur simultaneously.

The *ecological* processes of biodiversity optimization are considered as optimization of communities and populations in the given environment without any genetic modifications: a) optimization of the structure of the community through its "self-assembly" from the available species pool; b) adjustment of parameters of populations (species) due to behavioral or ontogenetic changes. The *microevolutionary* processes are considered as adjustment of intrapopulation diversity (width of realized population niche) due to changes in genetic diversity in the populations or changes in the average width of the norm of reaction in the population. During the *evolutionary* processes optimal parameters change as a result of evolution of species characteristics (e.g., the resource expenses, spectrum of consumed resources, mortality, fertility, etc.) and evolution of communities.

The general scheme of formation of diversity in accordance with optimum bio-diversity hypothesis is shown in Fig. 10. The process of formation of diversity consists of the following phases (the numbers correspond to the notations in Fig. 10):

1 – identification of optimal parameters of licenses, depending on environmental conditions and the evolutionary level of biota;

2 - filling licenses with species from the regional pool during "niche" and "neutral" interspecific interactions;

3 – adjustment of intrapopulation and intraspecific diversity due to behavioral changes and ontogenetic modifications of organisms;

4 - adjustment of optimum parameters of niches during succession;

5 - adjustment of intrapopulation and intraspecific diversity in the microevolution processes;

6 – changes in the optimal parameters of licenses during the evolution of organisms and communities.

Realized diversity substantially affects the efficiency of the functioning of populations and communities, including their environment-forming functions (in the concept of ecosystem services this functions are partially included in the groups of regulating and supporting services). Therefore, the scheme is closed to a kind of cycle due to dependence of the environment parameters on ecosystem functioning and biodiversity.

Can biosystems achieve optimum diversity? It seems that such cases can be quite rare. Optimal values of diversity depend on characteristics of the environment and biosystems, which are constantly changing. Therefore, the optimal values of diversity are also constantly shifting. At each time the vector of development of a biosystem is directed towards the optimal diversity, so we can say that the biosystem is in constant "pursuit" for its optimal state. We can assume that the undisturbed climax communities and their constituent populations (rather, coenopopulations) are the most close to the optimum values of diversity. How close are the parameters of actual biosystems to optimal values? It depends on the ratio of the rate of change of optimal values and the rate of adaptation of biosystems. Hereinafter, saying "optimal", we mean "closest to the optimal" because optimal biosystems can be quite a rare event in a constantly changing environment. However, cases when system not only achieve, but also support the optimum state for a long time are probably possible – such cases can be compared with the long-term evolutionary stasis.

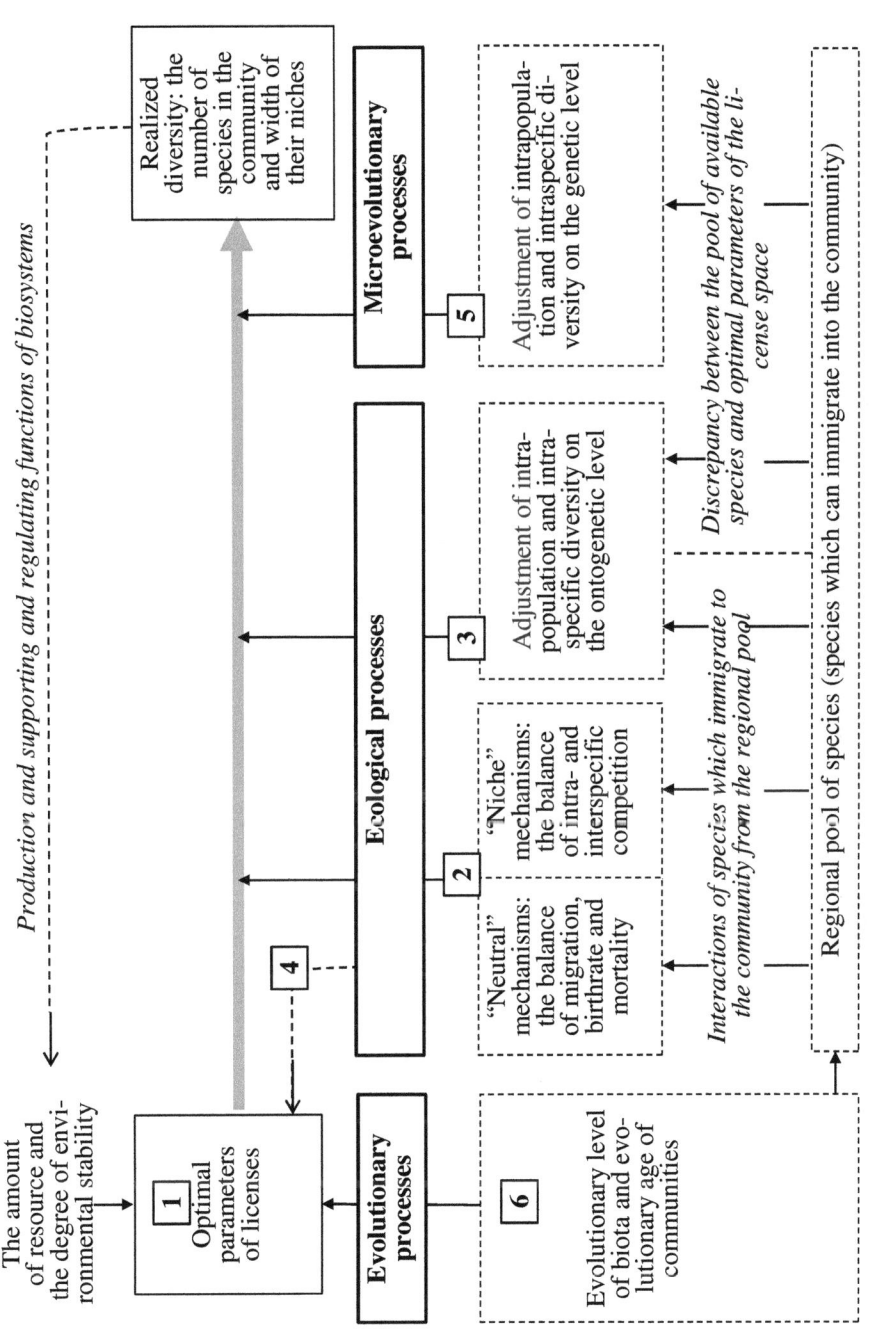

Figure. 10. The general scheme of the formation of biodiversity

5.2.1. Optimization of biodiversity in the ecological and microevolutionary processes

5.2.1.1.   Changes in the optimal diversity values during succession

Presented in this paper models consider the stationary states of biosystems and not the dynamics of their development. Despite this, the assumptions of the optimal biodiversity principle can be analyzed in terms of ecological successions comparing serial and climax stages with communities adapted to different environmental conditions. Further reasoning is based on the concept of succession of E. Odum.

Basic driver of succession is mismatch between volume of total community production and total community respiration. In the course of succession ratio of these parameters tends to "1", i.e., to a state when all assimilated energy is spent on maintenance of biomass. The ratio of total biomass to the assimilated energy increases while the biomass grows, and in climax each unit of absorbed energy supports the maximum biomass. This direction of development corresponds to the optimization criterion in the above models. Thus, succession can be compared with the optimization of communities. Optimization of diversity may be an additional factor of change of successional stages.

In accordance with the E. Odum concept, in the course of succession cycles of elements become more closed, stock volume and turnover time of nutrients increase, the stability of the whole system increases. Compared with the initial succession stages climax communities are more autonomous from the environment and have the active function of modification of the environment. In the course of succession the internal environment of the community becomes more stable, therefore, in accordance with the principle of optimum biodiversity the optimal values of species diversity increase and optimal values of intrapopulation diversity decrease. In other words, the cells of license space become smaller, and their number grows. Serial stages can be compared with communities adapted to less stable environmental conditions, and climax stages - with communities, adapted to more stable conditions. Transitions between stages form the optimal trajectory of succession (Fig. 11). Deviations from the optimal course of succession may be due to external influence on environmental conditions (e.g. pollution) or some transformations of the community (e.g. introduction of alien species).

Increase in the number of species and complexity of the community structure is accompanied by the improvement of mechanisms of regulation and stabilization of internal environment, which in turn leads to further increase in the number of species, and so on. This "autocatalytic" process goes up until the limit of partition of niches, which is determined by the amount of resource, the degree of stability of the environment and capabilities of species in the regional pool.

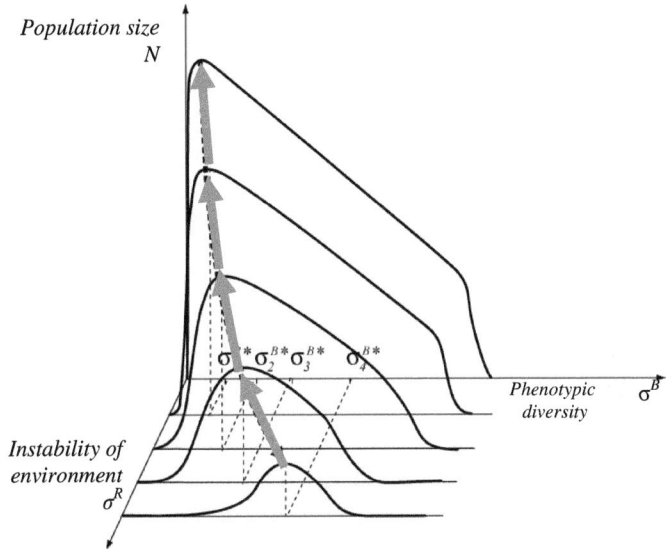

Figure 11. *Expected changes of the optimal parameters of populations in communities of different successional stages. Transitions between stages are shown by gray arrows.*

5.2.1.2.   *The combination of optimization with niche and neutral mechanisms of species interactions*

It is believed that ecological communities are formed by two main groups of mechanisms: a) the "niche" mechanisms, that are based on competition and division of ecological niches and b) the "neutral" mechanisms, that are based on the ratio of rates of migration, reproduction and mortality of different species. Inclusion in this picture of the optimal diversity values allows us to determine the possible range of action of each group. Predominance of neutral or niche mechanisms is determined by the relation of "richness" and stability of environment with the optimal geometry of license space, i.e. by the ratios D/D^* and R/R^*, where R is the total amount of available resource, R^* - the amount of resource necessary for the population with the optimal niche width, D – total range of the resource parameter (as in above model of community with the possibility of niche divergence, section 2.3), D^* - the optimal width of ecological niche (Bukvareva, Aleshchenko, 2013, Fig. 12). In other words, predominance of one or other group of mechanisms depends on the ratio of numbers of the optimal parts on which can be divided the amount of resource and the range of its parameter.

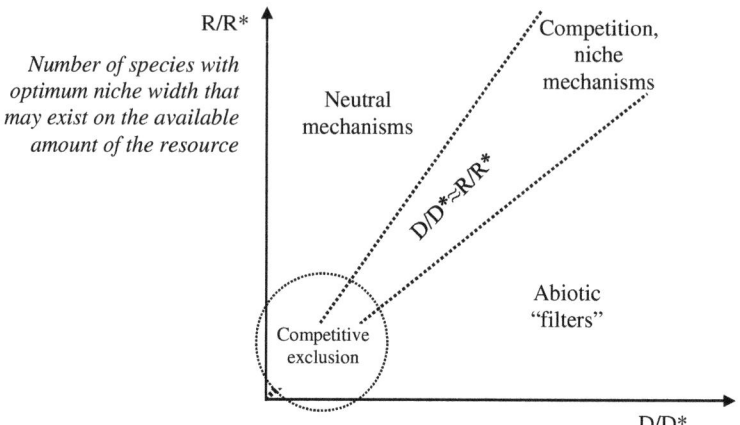

The number of niches with optimum width that may exist on
the available range of resource parameter

Figure 12. *Zones of priority action of basic mechanisms of community forming depending on the ratio of the number of optimal niches and number of species that may exist in the given environment.*

Based on the principle of optimal biodiversity we can assume the following scheme of the combined effect of different mechanisms of forming of community structure:

1 – the optimal values of species and intrapopulation diversity define number and width of licenses in accordance with amount of available resource, degree of environmental stability and evolutionary level of biota;

2 – licenses are filled during the interaction of species which immigrate from the regional pool; predominance of neutral or niche mechanisms is defined by the ratio between the amount and range of available resource with optimal parameters of the license space:

- competition and niche mechanisms work predominantly if the optimal number of niches which may exist in the available *range* of resource and the number of species that may exist on the available *amount* of resource are about the same $D/D^* \approx R/R^*$ (Fig. 12);

- neutral mechanisms work predominantly in a very "rich" environment where a great *amount* of resource allows the existence of a much larger number of species than the available *range* of resource allows (zone above the diagonal $D/D^* \approx R/R^*$);

- "abiotic filters" operate mainly in barren (harsh) environment where a small *amount* of resource allows the existence of a much smaller number of species than the available *range* of resource allows (area below the diagonal $D/D^* \approx R/R^*$).

Thus, the principle of optimal biodiversity suggests that the separation of niches is not obligatory condition for the formation of species diversity. Optimization of diversity may be an additional explanation for the known cases of coexistence of species almost in one niche that are hardly explained by the theory of competition. In the proposed scheme, the number of species in the community, and the width of their niches are determined primarily by optimal values of diversity, competition and divergence of species niches modify the community structure depending on the environmental conditions, the characteristics of regional species pools and community development stage.

5.2.1.3.   *Sympatric intraspecific forms as a mechanism for diversity optimization*

Intraspecific sympatric forms with different ecological characteristics usually found on islands, in lakes and in extreme habitats where species richness is depleted. Sometimes, however, intraspecific forms can also be observed under conditions without obvious signs of depletion of species richness.

We hypothesize that sympatric species subdivision may play the role of one of the mechanisms of diversity optimization. When forming several sympatric intraspecific forms that differ ecologically, the species occupies not one, but several subniches. Why doesn't form one large species niche, but instead the set of sympatric subniches form? As mentioned above, a very wide niche requires additional resources. The species can't be effective on a very wide range of environmental conditions due to contradictions between adaptations to different conditions. Therefore, the formation of several subniches instead of one wide niche can be interpreted as optimization of intraspecific diversity in those cases where the spectrum of available resources exceeds the optimal width of the ecological niche. The formation of intraspecific and intrapopulation sympatric forms brings intraspecific diversity closer to the optimum values, and at the same time allows to save wide range of resources and conditions, that is, increases the efficiency and stability of the species. At the same time, the formation of sympatric intraspecific forms can be interpreted as optimization of diversity at the community level. Development of sympatric intraspecific ecotypes can be regarded as a case when one species occupies multiple licenses (cells of license space) that were empty due to lack of species in the regional pool. That is, efficiency and sustainability of community may increase owing to formation of sympatric intraspecific forms.

In accordance with the principle of optimum diversity, the vector of optimization of species diversity in communities is directed towards reducing species number when destabilization of the environment and/or decline in resource amount, and it is directed towards increasing species number when stabilization of the environment and/or enlargement of resource amount. Problems do not arise if the regional pool has sufficient species and the considered habitat can freely exchange species with the surrounding habitats (the upper part of Fig. 13).

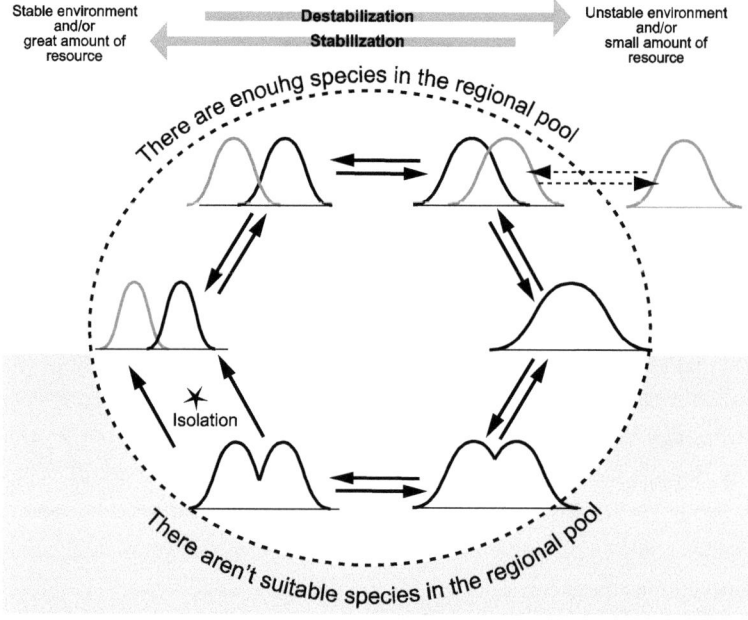

Figure 13. *The dynamic cycle of optimization of community diversity with free exchange with the regional species pool (the upper part) and with a lack of such exchange (the lower part).*

The most interesting case is when the community is isolated from surrounding habitats or the regional species pool is not enough to achieve optimal diversity. In this case, optimization can occur due to development of sympatric intraspecific forms. Sympatric intraspecific forms can be represented as a dynamic system, which constantly adjusts their diversity (width and number of subniches) in accordance with changes in the environment - from a large number of more specialized ecological forms - to fewer number of less specialized forms (the lower part of Fig. 13). Adaptive adjustment of the complex of intraspecific forms is possible until the isolation of forms from each other and the formation of new species.

The probability of formation of discrete ecological forms increases when the environment became more stable and/or more "rich" if there is a lack of regional species pool (in particular, if this habitat is isolated from other similar habitats). Isolation intraspecific forms and their splitting into separate species are also the most likely to occur when the environment becomes more stable.

5.2.1.4. Mechanisms of diversity optimization in populations and communities

Mechanisms of biodiversity optimization can be divided into two main groups:
- mechanisms which allow biological systems to achieve the optimal diversity (strictly speaking, mechanisms of optimization);
- mechanisms for realizing the advantages of the optimal biosystems in comparison with the suboptimal biosystems.

Mechanisms of the first group are well known to biologists (Table 2) and were briefly discussed above.

Table 2. *Mechanisms of biodiversity optimization in the ecological and microevolutionary processes*

Optimized parameter	Ecological processes	Microevolutionary processes
Intrapopulation diversity (width of ecological niche)	- Changes in behavior of individuals - Modification polymorphism (one genotype produces different phenotypes)	- Selection on the width of the reaction norm in the population - Changes in genetic diversity in the population
Diversity of intraspecific forms (the number of specific subniches in the community)	- Realization of different ontogeny trajectories of individuals in the population	- Disruptive selection
Number of species in the community	- Niche and "neutral" mechanisms of species interactions - Abiotic "filters"	- Sympatric speciation

Mechanisms of the second group are not clear and can be discussed only in the form of speculative assumptions. In accordance with our hypothesis, these mechanisms are based on the fact that populations and communities with the optimal diversity achieve greater abundance and biomass than suboptimal systems. We can assume that this gives them a better chance of survival and spread. At the level of organisms, such a mechanism is well known – it is the natural selection, i.e. in terms of our hypothesis the most likely survival and reproduction of individuals which have inner diversity the most close to the optimal values. However at the levels above organisms it is quite difficult to imagine such "selection" of the optimal populations, species and communities.

As a hypothesis, we can assume that the optimal parameters spread in populations and communities through the preferential distribution of their subunits, which have diversity the most close to the optimal values. At the *population level*, such subunits may be a subpopulation, a group of individuals or even single individuals. The optimal values of intrapopulation diversity can spread to a large subdivided population with individuals who are carriers of the optimal diversity values. Another mechanism may be a primarily expansion of subpopulations with optimal parameters inside the large subdivided population. It is not clear what can be a driving force of optimization of internal diversity of not-subdivided isolated population. At the *community level* the most likely mechanism for the spread of the optimal communities may be processes of succession type. In accordance with the proposed principle, community with the optimal diversity uses resources more effectively than suboptimal ones, that is, it can be regarded as the next stage of succession. Mechanisms of spreading of the optimal community are the same as the mechanisms of substitution serial communities during succession.

5.2.2. Changes in optimal values of diversity in the evolutionary processes

In the course of evolutionary processes the optimal diversity values change themselves (in contrast, during ecological and microevolution processes biosystems tend to adjust their parameters to achieve the optimal values). The optimal diversity values change because of increasing evolutionary level of organisms and the evolution of ecological communities. These processes are inextricably linked, but for clarity we consider them separately.

5.2.2.1. *Changes in the optimal diversity values with increasing evolutionary level of organisms*

The results of modeling (3.5) suggest that the increase in evolutionary level of organisms is accompanied by the following trends:

- increasing efficiency of populations, that is, reduction resource costs of a unit of biomass and increasing population size at a constant amount of the resource;

- extension of the zone of individual tolerance, which leads to lower values of the optimal intrapopulation diversity; the mechanisms of response to environmental changes move from population to individual level and become faster and more efficient;

- increasing the optimal values of species diversity.

Thus, increasing evolutionary level of organisms changes the optimal values of intrapopulation and species diversity, i.e., modifies the structure of the license space – cells became narrower, their number grows.

The historical example of taxonomic diversity growth in communities of evolutionary "advanced" organisms is the phased increase in marine diversity (Cambrian, Ordovician-Permian and Meso-Cainozoic), accompanied by an increment of proportion of mobile and "physiologically buffered" animals, i.e. animals with active forms of resource using, more effective internal regulation, less dependent on environment (Марков, Коротаев, 2007, 2008). The similar trends are observed for terrestrial biota.

The important factor of diversity growth in the evolution is increase in available energy flow through progressive evolutionary innovations, which allows organisms to expand the range of the usable resources and increase the intensity of their consumption. Modern theoretical ideas and empirical data suggest that enlargement of the amount of available energy is one of the major factors of biodiversity growth. This trend is fully consistent with predictions of the principle of optimum diversity.

In modern conditions, comparison of northern and tropical communities indirectly confirms our hypothesis. Arctic communities with low species diversity include greater proportion of primitive and archaic taxa than highly diverse tropical communities. It is obvious that the main factor of reducing species diversity in the Arctic and in other "harsh" habitats is small amount of available resources. Increase in proportion of primitive taxa in such habitats can be attributed to the success of "passive" forms of existence in harsh environment. However, according to Yu. Chernov (Чернов, 1991, 2002), additional reason for decline in species diversity in the north may be just increase in the proportion of archaic taxa.

Thus, we can assume that the principle of optimal diversity can be one of the additional explanations of general trend of increase in species diversity in communities of more evolutionary advanced organisms.

5.2.2.2. *Changes in the optimal diversity values in the course of evolution of communities*

In accordance with the concept of biocenotic regulation of evolution of Zherikhin (Жерихин, 1987), biocenosis have a stabilizing effect on species and limit the speed of their evolution during the "coherent" phases of slow evolution, when stable communities change gradually over long periods. During "incoherent" phases of desintegration of communities stabilizing influence of biocenosis is disabled and the rate of evolution increases. At the first glance, our results contradict this concept, predicting the increase in probability of speciation when the environment becames more stable. In fact, there is no contradiction, since in our model the tendency to discretization of intraspecific forms and speciation appears *not just in a stable* environment but *when increasing the stability* of environment, transition from an unstable environment - towards a more stable. Such phase begins after the destruction of communities, when the process of forming their "new generation" begin and the environment gradually becomes more stable. Substantial initial part of this process occurs in a relatively unstable environment during "incoherent" evolutionary phases.

Fig. 14 shows the changes in the optimal values of species and intrapopulation diversity during biocenotic crisis. Some internal or external factors cause the process of destruction of communities. The result is a sharp destabilization of biocenotic environment. After that, it remains relatively unstable for quite long time, until formation of new generation of sustainable communities. Destabilization of the environment causes a decrease in the optimal values of species diversity and increase in optimal values of intrapopulation diversity, licenses greatly expand, their number becomes smaller (gray "honeycomb" in Fig. 14). Optimization of diversity at this stage can occur, among other mechanisms, due to the transition of communities to earlier successional (truncation of succession, Жерихин, 2003), as they have the characteristics of adaptation to less stable environment. Then in the process of development and stabilization of new communities the optimal values of species diversity gradually increase and the optimal values of intrapopulation diversity decrease. Licenses become narrower and their number grows. If evolutionary level of organisms increases, licenses in the new communities become even smaller than they were before the crisis. At last this process reaches a plateau, on which the limit of fragmentation of licenses is determined by evolutionary level organisms, the amount of available resources and the degree of stability of the environment. Thus, in accordance with the principle of optimum diversity, during destruction and destabilization of communities intrapopulation diversity increases and thus provides a material for future forming of intraspecific forms and speciation. During the development of "new generation" of communities and their gradual stabilization intraspecific forms arise and become more discrete. This process can be finished with speciation if any form of reproductive isolation between forms arises.

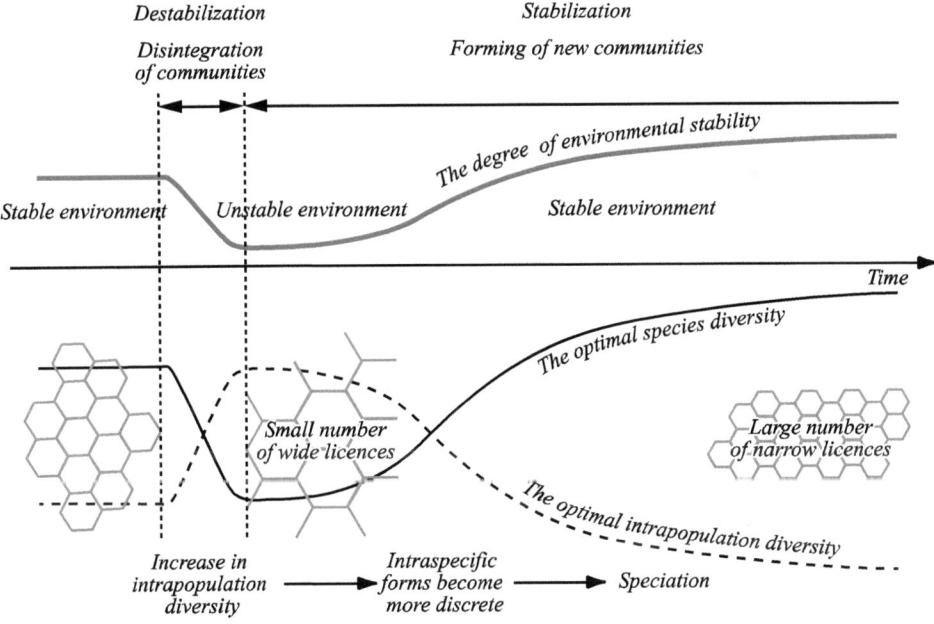

Figure 14. *Changes in the optimal values of species and intrapopulation diversity during biocenotic crisis and the subsequent development of new communities. Gray " honeycomb" shows the geometry of the license space.*

As mentioned above, the optimal diversity values in populations and communities response to changes of environmental stability in opposite manner, that in long time intervals can be expressed in antiphase oscillations of diversity in these hierarchical levels. The *periods of stable environment* are characterized by reduction of intrapopulation diversity, narrowing niches, discretization of intraspecific forms, speciation, increasing the number of species and complication of the hierarchical structure of communities. The *periods of unstable environment* are characterized by growth of intrapopulation diversity, expanding niches, extinction specialized species, decline of species diversity. These processes generally conform to the concept of coherent and incoherent evolution stages (Красилов, 1986). Thus, the optimization of diversity may be an additional factor in changing of diversity during evolution.

6.   The general scheme for the formation of biological diversity

Thus, the optimization can be considered as an additional mechanism for the formation of biodiversity in the ecological, microevolutionary and evolutionary processes (Fig. 15, paragraphs below correspond to the letters in the diagram).

A) The optimal parameters of the license-niche space (optimal values of intrapopulation and species diversity, i.e. the optimal width of niches and their number) depend on the evolutionary level of biota and environmental characteristics - degree of stability and amount of available resource. The optimal diversity values in populations and communities react to changes in the degree of stability in the opposite way: in a more stable environment the number of species increases, but intrapopulation diversity (niche width) reduces. That is, in a more stable environment the number of licenses grows and they become narrower. On the basis of the opposite reaction of the optimal diversity values in populations and communities we can make an assumption about the different role of diversity at these two levels: intrapopulation diversity is the basis for the adaptation to instability of environment and species diversity provides more efficient use of resources by differentiating of niches.

B) During the ecological processes license space is filled in the course of interaction of species which immigrate into the habitat from the regional pool. Predominance of neutral or niche mechanisms determined by the ratio of "richness" and stability of environment with optimal parameters of the license space:

- competition most strongly affects the formation of communities if the optimal number of licenses that are placed on the existing range of resource is approximately equal to the optimal number of species that may exist on the available amount of resource;

- neutral mechanisms work in a very "rich" environments where great amount of resource permits the existence of a much larger number of species than the optimal number of licenses that can be placed on a range of resource;

- "abiotic filters" act in meager or harsh environments where the number of species that may exist on the available resource significantly less than the optimum number of licenses that can be placed on a range of resource.

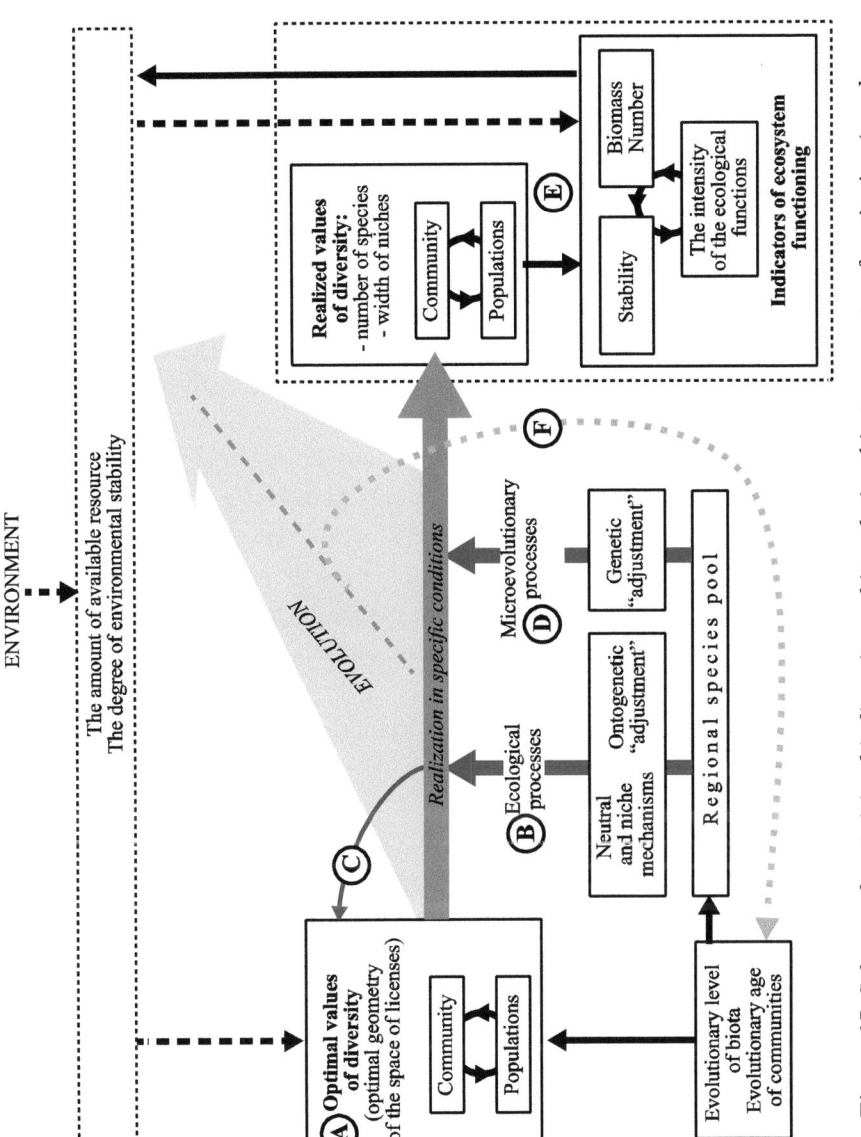

Figure 15. *Scheme of optimizing biodiversity and its relationship to ecosystem functioning (numbers correspond to paragraphs in the text).*

Optimal values of intrapopulation and species diversity arise independently of niche separation. Therefore, optimization can be the addition mechanism to niche separation and can explain the existence of sympatric ecological sibling species.

Optimization of width of population niches during the ecological processes occurs due to changes in the behavior of individuals and the modification polymorphism (one genotype produces diversity of phenotypes).

C) In the course of succession biocenosis develop mechanisms of homeostasis, so the optimal diversity values change - the number of species increases, niches become narrower. Initial succession stages can be compared with communities which are optimal in less stable environments, climax stages can be compared with communities which are optimal in more stable environments. Direction of succession towards maximizing biomass for the same amount of resource can be compared with the process of community optimizing. That is, the optimization of diversity may be an additional factor of substitution of succession stages.

D) If species from the regional pool are not enough to achieve the optimal diversity, intrapopulation and intraspecific diversity may be adjusted due to microevolutionary processes - selection of the width of the reaction norm, disruptive selection, increasing genetic diversity within populations. Formation of sympatric intraspecific forms can be considered as the mechanism of optimization of community structure and intrapopulation diversity when environment becomes more stable and available regional pool has not enough species.

E) Realized values of diversity define effectiveness of populations and communities, including total biomass that they can maintain per a unit of available resource. If diversity moves away from the optimal values due to anthropogenic disturbances or natural environmental changes the effectiveness of populations and communities decreases. Achieved efficiency of functioning, in turn, affects the ability to regulate the environment and thus, amount and stability of available resource, and through these indicators affects the optimal values of diversity. Thus, the relationship of "biodiversity - functioning" is bidirectional. Hypothesis ша epy optimal diversity can be used in the practical researches of relationship "biodiversity - ecosystem functioning" (Bukvareva, Aleshchenko, 2012).

F) During the evolution of species and communities the optimum values of diversity (the optimal geometry of license space) change. In particular, the growth of evolutionary level of organisms and progress in their autonomy from the environment lead to a decrease in the optimal values of intrapopulation diversity and increase in the optimal values of species diversity. It may be an additional factor of the growth of taxonomic diversity during evolution. Increase in evolutionary age of communities leads to growth of the optimal number of species and narrowing of niches. Antiphase oscillation of the optimal values of intraspecific and species diversity can be an additional explanation for the changes in biodiversity on coherent and incoherent evolution stages.

7. The principle of optimal diversity and strategy for the conservation and use of biodiversity

7.1. Transition biosystems in suboptimal state under anthropogenic impacts

As a result of anthropogenic impacts the optimal levels of biodiversity can be broken in two main ways: due to anthropogenic changes in the environment and because of the disruption of the structure of biosystems.

The most common anthropogenic *changes of the environment* are the destabilization and enrichment (e.g., fertilization, eutrophication), destabilization (e.g., disturbance of animals by humans), removal of biomass and destabilization of communities (e.g., logging, fishing). The main direction of adaptation of biosystems to these changes - increase intra-population diversity in response to the destabilization of the environment.

Anthropogenic impact on the *structure of biological systems* is expressed primarily in the reduction of the number of species and intra-population diversity when biosystems leave their optimal state. A typical example is reduction of intrapopulation diversity as a result of exploitation or habitat fragmentation. As indicated above, adaptation of populations to anthropogenic destabilization of the environment requires increasing intra-population diversity, but the main human impact on species and populations is the reduction of their numbers, intrapopulation and intraspecific diversity. Thus, wild species and populations are deprived of the opportunity to adapt to anthropogenic pressure. So adaptation mechanisms at the community level start work and typical species are replaced by others. In anthropogenic conditions there is a shift in community structure from K-strategists to r-strategists and from specialists to generalists. This shift corresponds to the modern proliferation of "gray" synanthropic biota.

Anthropogenic changes in the structure of biocenosis are usually expressed in the reduction in species diversity, when they also leave the optimal state.

7.2. The aims of management of ecosystem functions and services taking into account the changes in biodiversity

Since ecosystem functioning is inextricably linked with biodiversity, the strategies for the use of ecosystem services must take into account the current state and possible changes in biodiversity. The use of different services requires different management aims (Table. 3).

Table 3: *Management aims for different ecosystem functions and corresponding biodiversity changes*

Functions	Management aims	Biodiversity changes
Production	Maximum sustainable removal of biomass (maximum sustainable yield)	Decrease in biodiversity
Environment-forming	Efficient and sustainable ecosystem functioning	Maintaining of the natural level of biodiversity
Information	Getting information from natural systems	

There is a contradiction between the aims of producing maximum sustainable yield and maintain environment-forming functions. Management strategies to achieve these goals are different. When using environment-forming and information functions management aims coincide with the maintenance of natural levels of biodiversity and biomass. Environmental functions are the most effective and sustainable in undisturbed climax communities. Information functions in most cases also are maximal in undisturbed natural complexes. That is, in these cases the aim is to maintain the natural biodiversity levels which are close to optimal values. But when using the production functions the goal is to maximize sustainable crop. It is contrary to the maintenance of natural levels of diversity. The high productivity of communities requires simplifying of its structure and reducing of diversity. For this purpose an early and middle succession stages are optimal.

At the level of exploited populations maximization of the sustainable yield means the maximum increase in mortality. This corresponds to a strong destabilization of the environment. In this case adaptive trends of biosystems are as follows: increase in intra-population diversity; reduction in species diversity; reduction of total biomass. Exploitative pressure on populations eliminates the possibility of first mechanism, leaving only the second and third, which are contrary to the management aim for environment-forming and information functions. Minimizing of population biomass leads to a reduction and destabilization of the flows of matter and energy going through them and reduces their ecosystem functions.

Harvesting of natural systems make sense only if the value of ecosystem services is comparable with the value of removed biomass, but this ratio is not a typical. It should be expected that in most cases the value of environmental functions is many times more than the benefits from bioproduction. In these cases the strategy of "maximum sustainable yield" significantly reduces the total "benefit" of biodiversity. The way out is the "ecosystem approach" which suggests that the amount and form of exploitation of populations and ecosystems for bioproduction are severely limited by the requirement to maintain the structure and functions of ecosystems, species and populations.

7.3. The optimal diversity is the criteria for selecting priorities in environmental policy

For a long time the attention of the global conservation community has been focused on the tropical countries with the highest species diversity, so-called «megadiversity countries». However, in accordance with the principle of optimal diversity the most effectively communities have no maximum but the optimum diversity. Diversity is the important adaptation of biosystems to environmental conditions. For example, in the North in a more severe and less stable environment the small values of species diversity is optimal which are compensated by high intraspecific and intrapopulation diversity that ensures the effectiveness of ecosystem functions. Therefore, northern ecosystems have much less species diversity than tropical ecosystems, but play a key role in the biosphere regulation.

The situation is similar at the regional level: for example, the species diversity of peatbog communities is significantly smaller than diversity of meadows or mixed forests, but their ecosystem functions is not less important.

Criteria of conservation value of biosystems should be no formal diversity indexes (e. g., number of species), but the degree of its conformity to natural optimal values of diversity which provide the most effective ecosystem functions.

Literature

Aleshchenko G., Bukvareva E. 2010. Two-Level Hierarchical Model of Optimal Biological Diversity // Biology Bulletin. V. 37. N. 1. P. 1–9.

Bukvareva E., Aleshchenko G. 2012. The Principle of Optimal Biodiversity and Ecosystem Functioning // International Journal of Ecosystem. V. 2. N. 4. P. 78–87.

Bukvareva E., Aleshchenko G. 2013. Optimization, Niche and Neutral Mechanisms in the Formation of Biodiversity // American Journal of Life Sciences V. 1. N. 4. P. 174-183.

Levich A.P., Alekseyev V.L. 1997. The entropy extremal principle in the ecology of communities: results and discussion // Biophysics. V. 42. N. 2. P. 525-532.

Lurie D., Valls J., Wagensberg J. 1983. Thermodynamic approach to biomass distribution in ecological systems // Bulletin of Mathematical Biology. V. 45. N. 5. P. 869-872.

Wagensberg J., Valls J. 1987. The [extended] maximum entropy formalism and the statistical structure of ecosystems // Bulletin of Mathematical Biology. V. 49. N. 5. P. 531-538.

Алтухов Ю.П. 2003. Генетические процессы в популяциях. М.: Академкнига. 431 с.

Алещенко Г.М., Букварева Е.Н. 1991. Модель фенотипического разнообразия популяции в случайной среде // Журн. общ. биол. Т.52. № 4. С. 499-508.

Букварева Е.Н., Алещенко Г.М. 1997. Схема усложнения биологической иерархии в случайной среде // Успехи современной биологии. Т. 117. Вып.1ю С. 18-32.

Букварева Е.Н., Алещенко Г.М. 2005. Принцип оптимального разнообразия биосистем // Успехи современной биологии. Т. 125. № 4. С. 337-348.

Букварева Е.Н., Алещенко Г.М. 2010. Оптимизация разнообразия надорганизменных систем как один из механизмов их развития в экологическом, микроэволюционном и эволюционном масштабах // Успехи современной биологии. Т. 130. № 2. С. 115-129.

Букварева Е.Н., Алещенко Г.М. 2013. Принцип оптимального разнообразия биосистем. М.: КМК-Товарищество научных изданий. 522 с. (http://biodiversity.ucoz.com/load/knigi/princip_optimalnogo_raznoobrazija_biosistem/3-1-0-7)

Жерихин В. В. 1987. Биоценотическая регуляция эволюции // Палеонтол. журн. № 1. С. 3-12.

Жерихин В. В. 2003. Усечение сукцессий: возможный механизм диверсификации биомов // Жерихин В.В. Избранные труды по палеоэкологии и филоценогенетике. М.: КМК. С. 173-187.

Красилов В. А. 1986. Нерешенные проблемы эволюции. Владивосток: ДВНЦ. 140 с.

Левченко В.Ф., Старобогатов Я.И. 1990. Сукцессионные изменения и эволюция экосистем (некоторые вопросы эволюционной экологии) // Журн. общ. биол. Т. 51. № 5. С. 619-631.

Марков А.В., Коротаев А.В. 2007. Динамика разнообразия фанерозойских морских животных со-ответствует модели гиперболического роста // Журн. общ. биол. Т. 68. № 1. С.1-12.

Марков А.В., Коротаев А.В. 2008. Гиперболический рост разнообразия морской и континентальной биот фанерозоя и эволюция сообществ // Журн. общ. биол. Т. 69. № 3. С. 175-194.

Старобогатов Я.И., Левченко В.Ф. 1993. Экоцентрическая концепция макроэволюции // Журн. общ. биол. Т. 53. № 4. С. 389-407.

Чернов Ю.И. 1991. Биологическое разнообразие: сущность и проблемы // Успехи совр. биол. Т. 111. № 4. С. 499 – 507.

Чернов Ю.И. 2002. Биота Арктики: таксономическое разнообразие // Зоол. журн. Т.81. № 12. С. 1411-1431.

Supplement: The formal description of the model

Mathematical models have been realized by Gleb Aleshchenko

An ecological community Ω consists of a set of populations ω_μ ($\mu = 1,2,...M$). This community exists in *a* stochastic environment, which is characterized by the intensity of resource flow, R, and the degree of instability, s^R. Each population receives a part, ρ_μ, of the overall resource, R.

The lower level: population

There are two ensembles — $S = \{s_i\}$ and $F = \{f_i\}$ ($i = 1,2,...I$), the elements of which are placed in one–one correspondence. The ensemble S represents the set of values of the environmental parameter '*s*', which determines the possibility of resource consumption; the ensemble F represents the set of phenotypes.

The probability-distribution function of the sth value of the environmental parameter $V(s,c^R)$ is defined on the ensemble S. This function satisfies the normalization conditions as follows:

$$\forall s \in S : V(s,c^R) \geq 0; \ \sum_{s \in S} V(s,c^R) = \rho_\mu \qquad (1)$$

In the task of the lower level, ρ_μ is constant and may hence be considered to be "1".

The environmental parameter possesses a new value at each moment of time according to the probability-distribution function $V(s, c^R)$. The dispersion, σ^R, of the distribution $V(s, c^R)$ characterizes the degree of environmental instability.

The population size equals $N(t)$ at each moment of time. All individuals are distributed among F phenotypes (phenotypic classes). The number of individuals in phenotype f equals $n(t, f)$, i.e. $N(t) = \Sigma n(t, f)$.

The main phenotypic feature is considered as the ability to reproduce when a certain value of the environmental parameter is reached. Each element f of the ensemble F (phenotype) corresponds to a unique element s of the ensemble S (value of environmental parameter), which is the most favourable for reproducing this phenotype.

When the value $s*$ is achieved, the most fitting phenotype $f*$ and the group of phenotypes around it reproduce. The function $A(f, s*, c^A)$ determines the fraction of individuals of each reproducing phenotype when the environmental value $s*$ is realised. This function is defined for all elements $s*\hat{I}S$ and satisfies the conditions

$$\forall f \in F, s* \in S : 0 \leq A(f,s*,c^A) \leq 1; A(f*,s*,c^A)=1 \qquad (2)$$

The distribution $A(f, s*, c^A)$ may be interpreted in two angles: the farther a given phenotype is from $f*$, the lower is the fertility of reproducing individuals; or, a smaller fraction of individuals reproduce. Dispersion of the distribution of reproducing phenotypes σ^A may be interpreted as the index of the width of the individual-tolerance zone. This index is analogous to the "intraphenotypic component" proposed by J. Roughgarden.

Each reproducing phenotype generates a progeny of different phenotypes. The progeny of phenotype f* is distributed among the phenotypes according to the function $B(q, f^*, c^B)$, $(" f^* \hat{I} F)$, which is defined on the ensemble F and satisfies the normalization conditions

$$\forall f^*, q \in F : \sum_{q \in F} B(q, f^*, c^B) = 1; B(q, f^*, c^B) \geq 0 \tag{3}$$

Thus, the value of $B(q, f^*, c^B)$ defines the fraction of phenotype q in a progeny of phenotype f^*. The vectors c^R, c^B and c^A found in (1)–(3) are the parameters of corresponding distributions.

Dispersion of progeny distribution σ^B serves as the most important parameter defining the level of phenotypic diversity, which is reproduced at each step of population development. Therefore, σ^B was used as an argument of investigated dependencies during the computational experiment.

Mortality is defined by an exponential function with a constant death-rate coefficient 'd' i.e. number of dead during the time interval Δt is equal to $N(t)d\Delta t$.

The birth rate is modelled using a logistic function with the birth coefficient $b(t)$, which monotonously decreases as the population number grows:

$$b(t, N) = b_{max}(\frac{1 - N(t)}{N_{max}}) \tag{4}$$

In (4) b_{max}, N_{max} are constants that define the maximum values of birth coefficient and population number, respectively.

The model assumes that reproduction occurs in discrete moments of time. At each step of modelling (t = 1,2...), the value of the environmental parameter $s*$ is defined using a randomizer and in accordance with the probability distribution of the environmental parameter values $V(s, c^R)$. For a derived value $s*$, the distribution of progeny is calculated according to distributions (2), (3) and function (4) as follows:

$$b(t, N) \sum_{f \in F} A(f, s^*, c^A) B(q, f, c^B) n(t, f)(\forall q \in F)$$

Number of dead at the tth step of modelling is defined by the values $n(t, q)$ $d(" q \hat{I} F)$.

Thus, dispersion of the total number of individuals among the various phenotypes at the beginning of the $(t+1)^{th}$ step of modelling is defined by the following expression:

$$n(t+1,q) = n(t,q) + b(t,N)\sum_{f \in F} A(f,s^*,c^A)B(q,f,c^B)n(t,f) - n(t,q)d \qquad (5)$$

The system (5) is the main system of recurrent equations determining the dynamics of population number and phenotypic distribution. Step-by-step analysis of (5) for the initial conditions of $n(0, q)$ is achieved using the statistical testing procedure.

Results of modelling of population number N(t) at stationary mode under variation of σ^B show the presence of an optimal value σ^{B*}, which corresponds to the maximum value of population number N*. Values N* and σ^{B*} depend on the degree of environmental instability σ^R. The task at this level may be formally defined as follows:

$$N^*(\sigma^R) = \max\{N(\sigma^R, \sigma^B)\}. \qquad (6)$$

The value of maximum population number (6) obviously depends on the amount of available resource, which was assumed as "1" at the lower level and equals ρ_μ for the task at an upper level.

The upper level: ecological community

As stated above, community Ω consists of a set of M subsystems (populations) ω_μ ($\mu=1,2,...M$).

Each subsystem ω_μ using its inner parameter σ^B_μ maximizes its quantity N_μ which is always lower than a maximum possible quantity of population N^* (6) because each population gets only a part of the total amount of resource

$$N^*_\mu (\rho_\mu, \sigma^R) = (N_\mu (\rho_\mu, \sigma^R, \sigma^B_\mu) \qquad (7)$$

where ρ_μ is a resource provided by Ω for ω_μ.

On the upper level, let us consider the task of minimization by system Ω of its expenses on the maintenance of its subsystem ω_μ provided by the consumption of all available resource R by these subsystems. At the same time, it is suggested that the system Ω has two free parameters: M, the number of subsystems (i.e. populations); N_μ, the quantity of a subsystem ω_μ.

The system of upper level (community) determines the number of subsystems M and shares resource R with each subsystem that is a part of it ρ_μ ($\sum_{\mu=1}^{M} \rho_\mu = R$); a goal function of system Ω considers the "wishes" of the subsystems about their optimal quantity (7).

The condition for the full processing of resource R can be written in the following way:

$$\sum_{\mu=1}^{M} \rho_{1\mu} N_{\mu} = R,$$ (8)

where $\rho_{1\mu}$ – is the resource amount processed by one individual of μth population. The goal function of the system Ω can be defined as

$$E = \sum_{\mu=1}^{M} \beta_{1\mu} N_{\mu} + \sum_{\mu=1}^{M} \varphi_{\mu}(N_{\mu}, N_{\mu}^{*})$$ (9)

where $\beta_{1\mu}$ represents the spending on maintenance of one individual from μ^{th} population; φ_{μ} is a "penalty" function, where "penalty" is taken for deviation from optimal quantity of μth population.

Thus, the task of the system can be formulated in the following way: minimize the goal function (9) if limitation (8) is realized.

In such a statement, a solution to the given problem is highly difficult. So, following the desire to get at least approximate estimation of the system and its subsystems' behavior, it is suggested to simplify the given task of the upper level.

Let each subsystem get an equal amount of resource $\rho = R/M$ and consequently establish optimal quantity $N^{*}(\rho, \sigma^{R})$. Along with this, the limitation (8) can be written as

$\rho MN = R,$ (10)

and the goal function (9) will be $E = M(\beta_1 N + \varphi(N, N^{*}))$, where N is a quantity of each population that upper system Ω wants to reach, and N^{*} is an optimal quantity of each population.

The penalty function is 0 when $N = N^*$ and increases when N deviates from N^*; so without breaking the integrity of the task statement and of the assumption that the solution will be in a quadratic vicinity of N^*, the function $\varphi(N, N^*)$ can be written as $\varphi(N, N^{*}) = \beta_2(N^{*}-N)^2$.

Thus, the goal function of the upper level can be written as

$E = M(\beta_1 N + \beta_2(N - N^{*}(\rho, \sigma^{R}))^2) \rightarrow min_{M, N}$ (11)

Now the task of functioning of the two-level system under consideration can be formulated in the following way: the lower level maximizes quantity

$N^{*}(R/M, \sigma^{R}) \rightarrow max_{sB} N(R/M, \sigma^{R}, \sigma^{B})$, and the upper level minimizes spending i.e. the goal function E (11) at limitation (10).

The solution to such a two-level task without using iterative procedures can be obtained only in the case of a known function $N^{*}(R/M, \sigma^{R})$. Preliminary investigations of the stochastic population model give us such information. However, for sim-

plicity of the solution of the given task, without loss of integrity, we suggest that N*
is a linear relative to its indexes, i.e.

$$N^* = \alpha_1 R/M - \alpha_2 \sigma^R \tag{12}$$

Applying expression (12) to (11), we will get the task of the upper level, and so
its solution can be found without any difficulties. In a given case, we get $M^* \sim R/\sigma^R$,
where M* is an optimal number of populations.

It should be noted that other equivalent statements of the task are possible for
both the upper and lower levels. For example, for the upper level, it is possible to use
limitation (8) or (10) as a goal function, and goal function (9) or (11) as a limitation.
It is easy to show that the solutions to equivalent tasks will be functionally equal.

The model of ecological community with possibility of niche divergence

A formal description of the model is as follows. Model community of one
trophic level consists of I populations. Each population consists of J phenotypes, defined by their ability to reproduce when specific value of resource parameter is realized. $N_{ij}(t)$ – is a number of individuals of i-th population and j-th phenotype at a time
t $(i \in I, j \in J, t=1,2,...)$. The set of recurrent balance equations has the following form:

$$N_{ij}(t+1) = N_{ij}(t) - N^-_{ij}(t) + N^+_{ij}(t) \tag{1}$$

$N^-_{ij}(t)$ is a number of individuals which died and $N^+_{ij}(t)$ is a number of individuals which born during the time interval $(t, t+1)$ in j-th phenotype of i-th population.

$$N^-_{ij}(t) = d_i N_{ij}(t) \tag{2}$$

$$N^+_{ij}(t) = r_{ij}\beta_{jj}N_{ij}(t) + r_{ij-1}\beta_{j-1j}N_{ij-1}(t) + r_{ij+1}\beta_{j+1j}N_{ij+1}(t) \tag{3}$$

In (2) and (3) d_i is the mortality rate for the i-th species, r_{ij} – the fertility rate for
the relevant phenotype in i-th population, β_{nm} – the proportion of individuals born in
the m-th phenotype and passed to the n-th phenotype. In accordance with the condition (3) individuals can only be born of this phenotype and phenotypes neighboring to
it. Parameters β_{nm} should be restricted with $\beta_{jj} + \beta_{j-1j} + \beta_{j+1j} = 1$

The coefficients $r_{ij}(t)$ are defined in the following way:

$$r_{ij}(t) = r_{max}(1 - K_j(t)/R_j(t))(1 - S(t)/R) \tag{4}$$

where r_{max} – is the maximum allowable increase ratio; $K_j(t) = \sum_i N_{ij}(t)$ – is the sum
of all individuals consuming the resource of the j-th value; $S(t) = \sum_j K_j(t)$ – is a total
number of individuals at time t; $R_j(t)$ – is a random variable that determines the
amount of the j-th resource, realized in time t; $R = \sum_j R_j(t) = const$ – the amount of resource with a random distribution between the types of individuals.